Editor
Lorin Klistoff, M.A.

Editorial Manager
Karen J. Goldfluss, M.S. Ed.

Editor in Chief
Sharon Coan, M.S. Ed.

Illustrator
Wendy Chang

Cover Artist
Jessica Orlando

Art Coordinator
Denice Adorno

Creative Director
Elayne Roberts

Imaging
Ralph Olmedo, Jr.

Product Manager
Phil Garcia

Publisher
Mary D. Smith, M.S. Ed.

How to Calculate Measurements

Grades 1–3

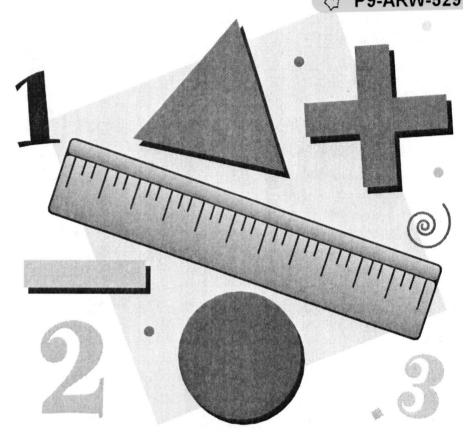

P9-ARW-529

Author

Mary Rosenberg

Teacher Created Resources

Teacher Created Resources, Inc.
6421 Industry Way
Westminster, CA 92683
www.teachercreated.com

ISBN-1-57690-952-2

©2000 Teacher Created Resources, Inc.
Reprinted, 2006
Made in U.S.A.

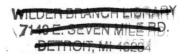

•••••••••••••••••• Table of Contents

A Note to Teachers and Parents

Welcome to the "How to" math series! You have chosen one of over two dozen books designed to give your children the information and practice they need to acquire important concepts in specific areas of math. The goal of the "How to" math books is to give children an extra boost as they work toward mastery of the math skills established by the National Council of Teachers of Mathematics (NCTM) and outlined in grade-level scope and sequence guidelines. The NCTM standards encourage children to learn basic math concepts and skills and apply them to new situations and to real-world events. The children learn to justify their solutions through the use of pictures, numbers, words, graphs, and diagrams.

The design of this book is intended to allow it to be used by teachers or parents for a variety of purposes and needs. Each of the units contains one or more "How to" pages and two or more practice pages. The "How to" section of each unit precedes the practice pages and provides needed information such as a concept or math rule review, important terms and formulas to remember, or step-by-step guidelines necessary for using the practice pages. While most "How to" pages are written for direct use by the children, in some lower-grade level books these pages are presented as instructional pages or direct lessons to be used by a teacher or parent prior to introducing the practice pages. In this book, the "How to" page details the concepts that will be covered in the pages that follow as well as how to teach the concepts. Many of the "How to" pages also include learning tips and extension ideas. The practice pages review and introduce new skills and provide opportunities for the children to apply the newly acquired skills. Each unit is sequential and builds upon the ideas covered in the previous unit(s).

About This Book

How to Calculate Measurements: Grades 1–3 presents a comprehensive overview of measurement for students at this level. It can be used to introduce and teach basic measurement to children with little or no background in the concepts. The units in this book can be used in whole-class instruction with the teacher or by a parent assisting his or her child through the book. This book also lends itself to use by a small group doing remedial or review work on measurement or for individuals and small groups in earlier grades engaged in enrichment or accelerated work. This book can also be used in a learning center that contains materials needed for each unit of instruction.

Children should be allowed to use a calculator to check computations. Some materials needed for this book include the following: ruler, yardstick, meterstick, thermometer, and protractor. Encourage children to use manipulatives to reinforce the concepts introduced in this book. Children should practice measuring or weighing objects whenever possible. Seize the moment and have children use an object such as a penny, a toy block, or a pencil to measure the number of units around a piece of paper, the dimensions of a desk, or the length and width of a door frame. Have children hold two objects and decide which weighs more.

If children have difficulty with a specific concept or unit within this book, review the material and allow them to redo the troublesome pages. It is preferable for children to find the work easy at first and to gradually advance to the more difficult concepts.

How to Calculate Measurements: Grades 1–3 highlights the use of various measuring devices and activities and emphasizes the development of proficiency in the use of basic measurement facts and processes for doing measuring. It provides a wide variety of instructional models and explanations for the gradual and thorough development of measuring concepts and processes.

The units in this book are designed to match the suggestions of the National Council of Teachers of Mathematics (NCTM). They strongly support the learning of measurement and other processes in the context of problem solving and real-world applications. Use every opportunity to have children apply these new skills in classroom situations and at home. This will reinforce the value of the skill as well as the process. *How to Calculate Measurements: Grades 1–3* matches a number of NCTM standards including the following main topics and specific features:

Problem Solving

The children develop and apply strategies to solve problems, verify and interpret results, are able to sort and classify objects, and solve word problems.

Communication

The children are able to communicate mathematical solutions through manipulatives, pictures, diagrams, numbers, and words. Children are able to relate everyday language to the language and symbols of math. Children have opportunities to read, write, discuss, and listen to math ideas.

Reasoning

Children make logical conclusions through interpreting graphs, patterns, and facts. The children are able to explain and justify their math solutions.

Connections

Children are able to apply math concepts and skills to other curricular areas and to the real world.

Number Sense and Numeration

Children learn to count, label, and sort collections, as well as learn the basic math operations of adding and subtracting.

Concepts of Whole Number Operations

Children develop an understanding for the operation (addition and subtraction) by modeling and discussing situations relating math language and the symbols of operation (+ and –) to the problem being discussed.

Other Standards

Children work toward **whole-number computation** mastery as they model, explain, and develop competency in basic facts, mental computation, and **estimation** techniques.

Children explore **geometry** and develop **spatial sense** by describing models, drawing and classifying shapes, and relating geometric ideas to number and measurement ideas.

Children learn about **statistics** and **probability** by collecting and organizing data into graphs, charts, and tables.

Children develop concepts of **fractions** through the use of pattern blocks.

1 ► How to ••••••• Use Nonstandard Measurement

Learning Notes

Nonstandard measurement is the use of items as measurement tools that are not uniform in size. Using fingers to measure something is an example of nonstandard measurement. One person's fingers are not necessarily the same size as another person's fingers.

In this unit, children practice measuring different objects using nonstandard measurement tools. In addition, they use both nonstandard and customary measurement tools to measure the same items and compare the two answers. Also in this unit, the children find the difference between measurements using palms and spans. They use nonstandard measurement tools to measure both distance and length.

Materials

- ruler labeled in inches

Teaching the Lesson

Before beginning the activities, discuss the different terms and objects used for measuring tools: a finger, a palm, a span, a person's foot, and the customary measurement called a *foot* (12 inches).

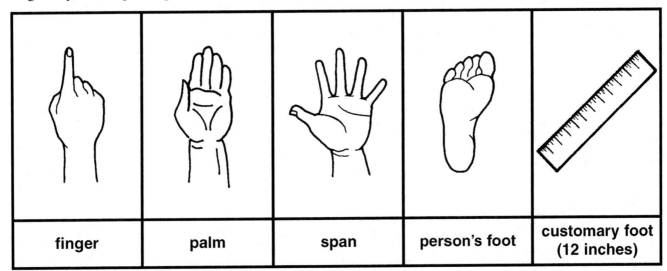

| finger | palm | span | person's foot | customary foot (12 inches) |

Model how to measure an item using nonstandard measurement. (*Note:* Remind the children to always begin measuring at the beginning of the object.)

Step 1	**Step 2**	**Step 3**
Start at the beginning of the object.	Measure to the end of the object.	Record the answer.

The mat is 4 spans long.

Step 1

Line up the index finger with the beginning edge of the item being measured.

Step 2

Use the index finger on your other hand to measure the next segment.

Step 3

"Leap frog" the first finger over the second finger. Then the second finger over the first finger until the entire length of the item has been measured. Record the number of fingers used.

Directions: Use your index fingers to measure the following items. Then have a friend use his or her index fingers to measure the same items.

Item	My Fingers	Friend's Fingers
1. chalk	_____ fingers	_____ fingers
2. marker	_____ fingers	_____ fingers
3. shoe	_____ fingers	_____ fingers
4. chalk eraser	_____ fingers	_____ fingers
5. stapler	_____ fingers	_____ fingers

6. Were your answers the same as your friend's? Why or why not?

_____

_____

7. If your fingers were bigger, would this change your answers? Why?

_____

_____

A hand can be used two ways to measure items. Having the hand open and fingers together is called "palm" measuring. A hand "span" has the hand open and the fingers spread apart.

palm

span

Directions: Place a hand at one edge of the item. Then place the other hand next to the first hand and "leap frog" the first hand over the second hand. Continue "leap frogging" until the width of the item has been measured. Record the number of hand palms or spans used.

Item	Palms	Spans	Difference (palms – spans)
1. desk	_____ palms	_____ spans	
2. table	_____ palms	_____ spans	
3. open newspaper	_____ palms	_____ spans	
4. chair	_____ palms	_____ spans	
5. radio	_____ palms	_____ spans	
6. door	_____ palms	_____ spans	
7. bookcase	_____ palms	_____ spans	
8. window	_____ palms	_____ spans	

9. What was the greatest difference?

_____

10. What was the smallest difference?

_____

The measurement called a *foot* is 12" long.

A person's foot can also be used to measure items.

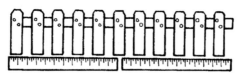 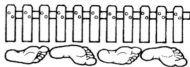

Directions: Line the heel of one foot up against the beginning of the object. Then walk heel-to-toe the length (or distance) of the object. Record the number of feet. Then measure the distance using a ruler. (Two rulers might be helpful.) Record the number of feet. Find the difference between measuring with your feet and a standard foot ruler.

Distance from your desk to:	Your Feet	Standard Foot	(Your Feet – Standard Feet) Difference
1. the door	_____ feet	_____ feet	
2. the trash basket	_____ feet	_____ feet	
3. the bookcase	_____ feet	_____ feet	
4. the teacher's chair	_____ feet	_____ feet	
5. the window	_____ feet	_____ feet	
6. the chalkboard	_____ feet	_____ feet	
7. the games area	_____ feet	_____ feet	

8. What was the greatest difference?

_____

9. What was the smallest difference?

_____

10. What would happen to the answers if your foot was longer than 12 inches?

_____

Learning Notes

Standard measurement involves using items as measurement tools that are uniform in size. Some examples include a set of dominoes that are all the same size, a deck of playing cards that are all the same size, and a box of paper clips that are all the same size.

In this unit, children estimate the length of various items in reference to standard measurement items. They measure items using standard measurement tools. In addition, children measure line segments using a standard measurement tool of their choice.

Materials

- buttons
- paper clips
- books
- rulers
- multilink cubes

- craft sticks
- clothespins
- crayons
- teddy bear counters
- other items of a standard size

Teaching the Lesson

Model how to measure items using standard measurement tools by lining up the measurement tool (paper clips) at the beginning of the item (pencil) being measured. Then continue lining up the paper clips until the end of the pencil is reached. Record the length of the pencil to the nearest paper clip.

The pencil is 4 paper clips long.

After children understand how to measure using standard measurement tools, discuss the concept of "estimating." An estimate is a "guess." Place a paintbrush and a teddy bear counter side by side. Have the children look at the size of the teddy bear counter and the length of the paintbrush. Have the children share "reasonable guesses" with you. Model how to record the estimate, then use the teddy bear counters to measure the length of the paintbrush, and record the answer. Compare the estimate to the actual number. Give the children several opportunities to practice this skill before completing the practice pages.

Estimate	Actual
4 teddy bears	3 teddy bears

Directions: Teddy bear counters can be used as a standard measurement, because they are all the same size. Look at the pictures below. Estimate the length of each item and write down the estimate. Then measure each item using the teddy bear counters and record the actual length to the nearest teddy bear counter. Look at the sample.

Estimate = 4 teddy bears   Actual = 3 teddy bears

Item to Measure	Estimate	Actual
1. pencil	_____ teddy bears	_____ teddy bears
2. glue bottle	_____ teddy bears	_____ teddy bears
3. journal	_____ teddy bears	_____ teddy bears
4. hand	_____ teddy bears	_____ teddy bears
5. scissors	_____ teddy bears	_____ teddy bears
6. stapler	_____ teddy bears	_____ teddy bears
7. shoe	_____ teddy bears	_____ teddy bears
8. book	_____ teddy bears	_____ teddy bears

Directions: Pick one of the following standard measurement items: paper clips, clothespins, buttons, or multilink cubes. Use the measurement item to measure the different line segments. Line up the measurement item below the beginning of the line segment. Measure to the end of the line segment. Record the measurement to the nearest measurement item. Look at the sample.

The line is 4 paper clips long.

1. •————————————————————•   The line is _____
   _____

2. •——————————•   The line is _____
   _____

3. •————————————————•   The line is _____
   _____

4. •————————•   The line is _____
   _____

5. •——————————————•   The line is _____
   _____

6. •——————————•   The line is _____
   _____

7. •————————————•   The line is _____
   _____

8. •————————————————————•   The line is _____
   _____

9. •———•   The line is _____
   _____

10. •——————————•   The line is _____
    _____

Standard measurement uses items that are uniform in size to measure the length of an object. When measuring objects with standard measurement items, line the measurement items up directly underneath the object. Make sure the beginning of the first measurement item is placed underneath the beginning of the object.

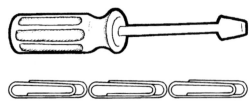

The screwdriver is 3 teddy bear counters long.

The screwdriver is 3 paper clips long.

Directions: Use teddy bear counters, multilink cubes, paper clips, plastic chains, and one other standard item of your choice to measure the following objects.

Item	Teddy Bear Counters	Multilink Cubes	Paper Clips	Plastic Chains	Other
1. envelope					
2. crayon box					
3. ruler					

Learning Notes

Customary measurement is one type of measurement standard that is common throughout the world. For example, a foot is a common standard that is used in the United States. People around the country know that a foot is 12 inches in length.

In this unit, children will use a ruler to measure objects to the nearest inch and will compare the length of an object to one foot (12 inches). The children will also complete a table that shows the relationship between inches, feet, and yards. The children will use this information to answer questions. In addition, rulers will be used to figure out the number of miles from one town to another.

Materials

- rulers
- paper

- yardsticks (or string cut into 3-foot lengths)
- scratch paper

Teaching the Lesson

Go over the markings on the ruler with the children.

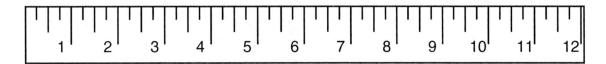

1 ruler = 12 inches

Model for the children how to use the ruler to measure and draw a line to a specified length on the paper. Monitor the children to make sure they are measuring from the beginning of the ruler—not from the middle of the ruler.

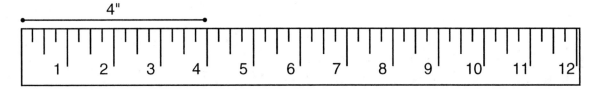

Go over the markings on a yardstick with the children. Point out that the yardstick is the length of 3 rulers. In other words, it is 3 feet long. Tell the children that it is divided into 36 inches. Give each child a yardstick or a piece of string 36 inches in length. Have the children practice measuring to different points in the room. For example, have the children measure from your chair to the classroom door.

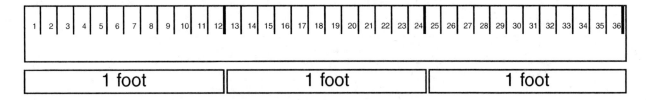

1 yard = 3 feet = 36 inches

When measuring with a ruler, always place the beginning of the ruler directly underneath the beginning of the object being measured. Measure to the end of the object and record the number.

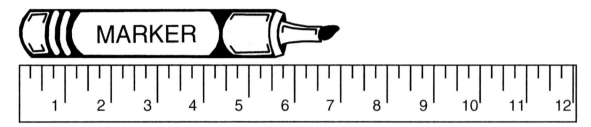

The marker is 7 inches long.

Directions: Measure each item to the nearest inch.

1.	_____ in.
2.	_____ in.
3.	_____ in.
4.	_____ in.

Directions: Use a ruler to measure the following actual (or real) body parts. Circle whether it is less than a foot (less than 12") or more than a foot (more than 12").

5.	less	more	**6.**	less	more
7.	less	more	**8.**	less	more
9.	less	more	**10.**	less	more

3 Practice • • • • • • • • • • • How Many Inches in a Foot?

Inches	6 in.	12 in.	18 in.	24 in.	30 in.	36 in.	42 in.
Feet	$\frac{1}{2}$ ft.	1 ft.	$1\frac{1}{2}$ ft.	2 ft.	$2\frac{1}{2}$ ft.	3 ft.	$3\frac{1}{2}$ ft.
Yards	$\frac{1}{6}$ yd.	$\frac{1}{3}$ yd.	$\frac{1}{2}$ yd.	$\frac{2}{3}$ yd.	$\frac{5}{6}$ yd.	1 yd.	$1\frac{1}{6}$ yd.

Directions: Each item was measured in either feet, yards, or inches. Use the chart above to help you complete the table of information for each item below.

Item	Inches	Feet	Yards
1. suitcase 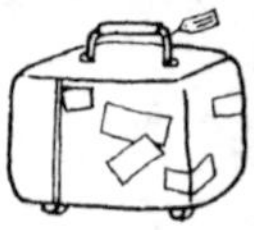	______ in.	______ ft.	$\frac{5}{6}$ yd.
2. dollar bill 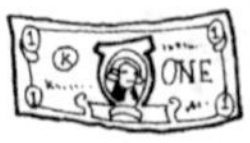	______ in.	$\frac{1}{2}$ ft.	______ yd.
3. trash can 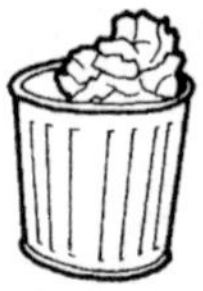	18 in.	______ ft.	______ yd.
4. dog 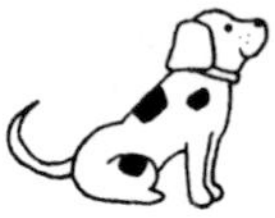	______ in.	3 ft.	______ yd.
5. toy car 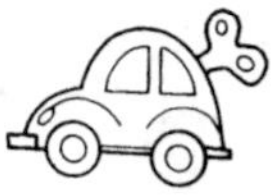	______ in.	______ ft.	$\frac{1}{3}$ yd.
6. baby 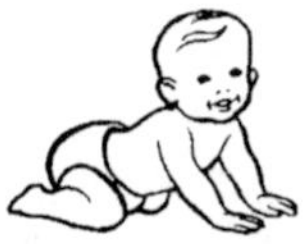	24 in.	______ ft.	______ yd.
7. ball 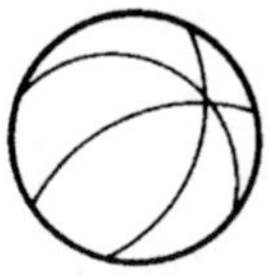	______ in.	______ ft.	$\frac{2}{3}$ yd.
8. bird house 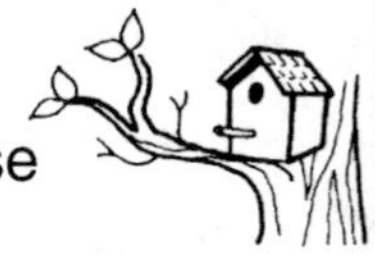	______ in.	$1\frac{1}{2}$ ft.	______ yd.

Directions: Figure out the length of each highway and record the number of miles on the map. Use the map to answer the questions.

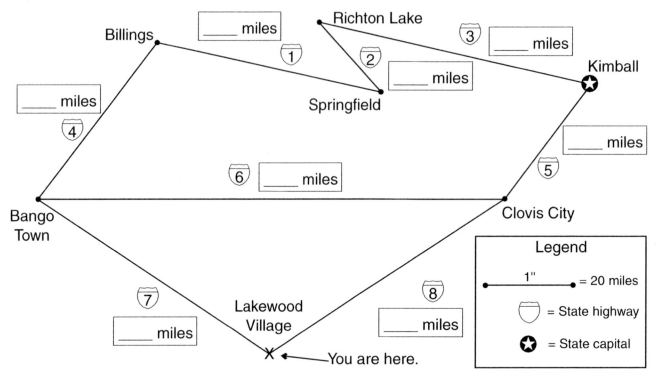

Directions: After recording the number of miles for each of the highways, use the map to answer the questions below.

How many miles is it from Lakewood Village to Springfield?

1. Taking ⑦ to ④ to ① = _____ miles.

2. Taking ⑧ to ⑤ to ③ to ② = _____ miles.

3. Which route is the longest? _____

4. What route would you take to get from Kimball to Bango Town? _____

5. How many miles is that route? _____

6. What route would you take to get from Clovis City to Billings? _____

7. How many miles is that route? _____

8. Which route would you take to get from Clovis City to Bango Town?

    _____

9. How many miles is that route? _____

10. If you can travel 150 miles a day, which cities would you visit, and how many miles would you have to drive to visit those cities? _____

    _____

Learning Notes

The *metric system* is one type of measurement standard that is common throughout the world. For example, a meter is a common standard that is known in most countries. People throughout the world know that a meter is 100 centimeters in length.

In this unit, children will measure objects to the nearest centimeter or meter. In addition, they will measure objects to find the perimeter.

Materials

- rulers labeled in centimeters
- metersticks

Teaching the Lesson

Introduce the metric ruler (enlarged illustration below) divided into segments, one centimeter in length. Go over the centimeter markings with the children.

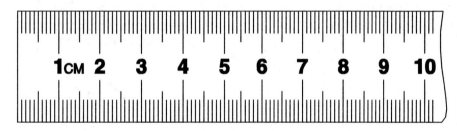

Model for the children how to measure an object using the ruler. First, place the beginning of the ruler directly under the beginning of the object. Next, measure to the end of the object. Record the length to the nearest centimeter.

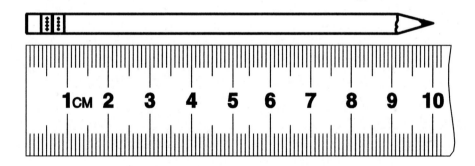

The pencil is 10 cm long.

Show the children a meter stick. Compare the length of the meter stick to the ruler labeled in centimeters. Tell them that one meter is 100 centimeters long.

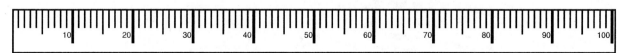

1 meter = 100 centimeters

Centimeters are used to measure items in the metric measuring system. One centimeter is less than half of an inch. To measure in centimeters, place the beginning of the centimeter ruler directly underneath the beginning of the object being measured. Measure to the end of the object and record the length in centimeters. The short way to write centimeters is "cm".

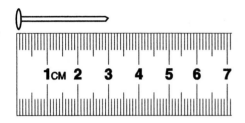

The nail is 3 centimeters or 3 cm long.

Directions: Measure each item below to the nearest centimeter. Write the answer on the line.

1.

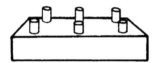

_____ cm

2.

_____ cm

3.

_____ cm

4.

_____ cm

5.

_____ cm

6.

_____ cm

7.

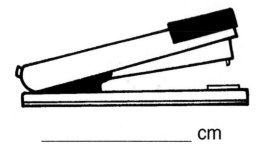

_____ cm

8.

_____ cm

4 ▶ Practice •••••••••••••••• Finding the Perimeter

Centimeters can be used to find the *perimeter* (the distance around an object). First, measure each side and record the number. Then, add the numbers together to find the perimeter.

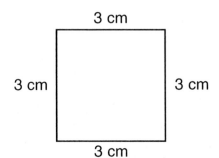

3 cm
3 cm
3 cm
3 cm

3 cm + 3 cm + 3 cm + 3 cm = 12 cm
The square has a perimeter of 12 cm.

Directions: Measure each side in centimeters and then find the perimeter for each shape.

1.

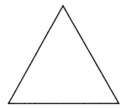

_____ + _____ + _____ = _____ cm

2.

_____ + _____ + _____ + _____ = _____ cm

3.

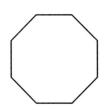

_____ + _____ + _____ + _____ + _____ +
_____ + _____ + _____ = _____ cm

4.

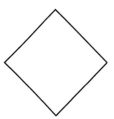

_____ + _____ + _____ + _____ = _____ cm

5.

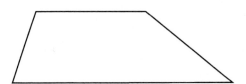

_____ + _____ + _____ + _____ = _____ cm

6.

_____ + _____ + _____ = _____ cm

A *meter* is the same as 100 centimeters. A meter is a little longer than 39 inches.

Use meters when measuring tall or long objects.

Use centimeters when measuring small objects.

Directions: Look at the pictures below. Decide whether to measure the actual item in meters (m) or centimeters (cm). Measure the item to the nearest centimeter or meter and record the answer.

Item	Centimeters	Meters
1. floppy disk	_____ cm	_____ m
2. penny	_____ cm	_____ m
3. door	_____ cm	_____ m
4. height of a friend	_____ cm	_____ m
5. book	_____ cm	_____ m
6. juice box	_____ cm	_____ m
7. chalkboard	_____ cm	_____ m

Learning Notes

In this unit, children will be given a chart showing the ratio among cups, pints, quarts, and gallons. They will develop an understanding of equivalent measures and will measure ingredients according to the recipes to make individual servings of cupcakes, muffins, and corn bread. The children will be introduced to the metric system's volume measurement of deciliters and liters. They will estimate the amount of water needed to fill different sizes of containers.

Materials

- measuring cups and spoons (both U.S. measurements and metric measurements)
- tubs
- sand
- craft sticks
- 7 oz. (198 g) waxy paper cups
- electric frying pan with lid
- bowls
- small container of egg substitute
- muffin and corn bread mix
- cake mix (frosting is optional)

Teaching the Lesson

Have the children practice measuring dry (sand) and wet (water) ingredients. Model for the children how to fill the measuring cup or spoon with the sand, and use the craft stick to level off the ingredients. When measuring wet ingredients, place the cup on a flat surface and stand at eye level in order to get an accurate reading of the wet ingredient.

Here are the steps to measure the ingredients:

Step 1	Step 2	Step 3
Fill the measuring spoon with the dry ingredient.	Holding the measuring spoon over the bowl full of mix, use the edge of the craft stick to level the measuring spoon. Let the excess mix fall back into the bowl.	Pour the spoonful of mix into the paper cup.

Give the children plenty of time to practice measuring ingredients and to experiment with ratios among the different measurements. For example, ask them, "How many tablespoons are in a ¼ cup?" or "How many liters does it take to fill the bowl with water?" Call on the children to model how many cups it takes to make a pint, how many cups to make a quart, etc. This will make the abstract concept more concrete for the children.

Allow 30–40 minutes to complete the cooking activities.

		1	2	4	8	16
	cups (c.)	1	2	4	8	16
	pints (pt.)	$\frac{1}{2}$	1	2	4	8
	quarts (qt.)	$\frac{1}{4}$	$\frac{1}{2}$	1	2	4
	gallons (gal.)	$\frac{1}{16}$	$\frac{1}{8}$	$\frac{1}{4}$	$\frac{1}{2}$	1

Directions: Draw a line matching each set of containers to its equivalent measurement. Use the chart above to help you answer the questions.

1.

2.

3.

4.

Directions: Use the chart at the top of this page to answer the following questions.

5. 1 gallon = _____ cups

6. 2 quarts = _____ pints

7. 1 cup = _____ pint

8. 8 pints = _____ gallon

9. 1 quart = _____ cups

10. 8 pints = _____ quarts

11. 8 cups = _____ gallon

12. 4 cups = _____ quart

Cupcakes

Materials

- cake mix
- tablespoon
- bowl of water
- bowl for mix
- small carton of egg substitute

- 1 craft stick for each person
- electric frying pan (with lid) set at 400° F
- one 7–ounce (198 g) waxy paper cup for each person
- frosting (optional)

Directions

In each paper cup, put 3 tablespoons (45 mL) of the cake mix, 1 tablespoon (15 mL) of water, and 1 tablespoon (15 mL) of the egg substitute. Stir the mixture thoroughly. (If the mixture is watery, add a small amount of the cake mix and continue stirring.) Throw the craft stick away, place the entire cup in the electric frying pan, and cover with the lid. (Depending upon the size of the frying pan, 9–18 cupcakes can be baked at one time.) Bake for about 15–20 minutes at 400° F or until a toothpick inserted into a cupcake comes out clean. Let the cups cool for a few minutes, then "peel" the cup and eat the cupcake! If desired, the cupcake can be frosted. Each cake mix makes approximately 10–14 cupcakes.

Muffins

Materials

- instant muffin or corn bread mix
- tablespoon
- bowl of water
- bowl for mix

- small carton of egg substitute
- 1 craft stick for each person
- electric frying pan (with lid) set at 400° F
- one 7–ounce (198 g) waxy paper cup for each person

Directions

In each cup, put 3 tablespoons (45 mL) of the muffin or corn bread mix, 1 tablespoon (15 mL) of water, and a small drizzle of the egg substitute. Stir the mixture thoroughly. (If the mixture is watery, add a small amount of dry mix to the cup and continue stirring.) Throw the craft stick away, place the entire cup in the electric frying pan, and cover with a lid. Bake for approximately 15 minutes at 400° F or until a toothpick inserted into the cup comes out clean. (Depending upon the size of the frying pan, 9–18 muffins can be baked at one time.) Let the cups cool for a few minutes, then "peel" the cup off and eat the tasty treat! Each mix makes approximately 7 servings.

One *liter* is about the same as 4 cups. One *deciliter* is a little bit less than half of a cup. It takes 10 deciliters to make 1 liter.

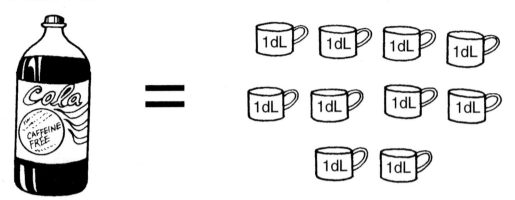

1 liter (L) = 10 deciliters (dL)

Directions: Figure out the best measurement to use for each of the containers. Circle whether it would be "dL" (deciliters) or "L" (liters).

1.	2.	3.	4.	5.
dL L	dL L	dL L	dL L	dL L

6.	7.	8.	9.	10.
dL L	dL L	dL L	dL L	dL L

11.	12.	13.	14.	15.
dL L	dL L	dL L	dL L	dL L

Learning Notes

In this unit, children will use tables showing the ratios between ounces and pounds, grams and kilograms, and pounds and kilograms. They will estimate whether an object should be weighed in ounces or pounds and whether an object should be weighed in grams or kilograms. The children will use a calculator to convert pounds to kilograms and kilograms to pounds.

Materials

- calculators
- bathroom scale
- balances with weights (or use teddy bear counters, beans, unifix or multilink cubes, etc.)
- items to weigh—fruit, crayons, paper towels, erasers, books, etc.

Teaching the Lesson

Introduce the concept of weight through modeling how to use the bathroom scale and the balance.

Ask the children which kind of scale is best for weighing heavy objects (bathroom scale) and which kind of scale is best for weighing light objects (balance).

Have the children practice weighing different objects with the balance. Then order the objects by weight, from lightest to heaviest.

When using the balance, follow the steps below.

Step 1

Place the item to be weighed (crayons) on one side of the balance.

Step 2

Put the weights (teddy bear counters) one at a time on the other side of the balance. Continue adding weights to the balance until both sides of the balance are at the same level.

Step 3

Add up the weights used to make the crayons balance.

Step 4

Record the weight of the crayons.

The crayons weighed 7 teddy bear counters.

Directions: There are 16 ounces in 1 pound. Complete the chart.

Ounces (oz.)	16	1.	48	2.	80	3.	112	4.	5.		160
Pounds (lbs.)	1	2	6.	4	7.	8.	9.	10.	9	11.	

Directions: Look at the objects below. Figure out the best measurement to use to weigh each object. Circle "oz." for ounces or "lb." for pounds.

12.	13.	14.	15.	16.
oz.　lb.	oz.　lb.	oz.　lb.	oz.　lb.	oz.　lb.

17.	18.	19.	20.	21.
oz.　lb.	oz.　lb.	oz.　lb.	oz.　lb.	oz.　lb.

22.	23.	24.	25.	26.
oz.　lb.	oz.　lb.	oz.　lb.	oz.　lb.	oz.　lb.

Challenge

Figure out how much you weigh in ounces by taking your weight in pounds and multiplying it by 16. How much do you weigh in ounces? _____ ounces

6 ▶ Practice ••••••••••••••• Grams and Kilograms

Directions: Complete the chart.

Grams (g)	1,000	2,000	1.	2.	5,000	3.	4.	8,000	5.	6.
Kilograms (kg)	7.	2	8.	9.	10.	6	11.	12.	13.	10

Directions: Look at the objects below. Would you use grams (g) or kilograms (kg) to weigh the object? Circle "g" for grams or "kg" for kilograms.

14. g   kg

15. g   kg

16. g   kg

17. g   kg

18. g   kg

19. g   kg

20. g   kg

21. g   kg

22. g   kg

23. g   kg

24. g   kg

25. g   kg

26. g   kg

27. g   kg

28. g   kg

Challenge

Use a calculator to figure out how much you weigh in kilograms by taking your weight in pounds and multiplying it by .454. How much do you weigh in kilograms? _____ kg

Pounds (lbs.)	1	2	3	4	5	6	7	8	9	10
Kilograms (kg)	.45	.90	1.35	1.80	2.25	2.70	3.15	3.60	4.05	4.50

Directions: Each animal's weight is shown in pounds (lbs.). Figure out each animal's weight in kilograms (kg) and write the answer in the box. Use the above chart to help you.

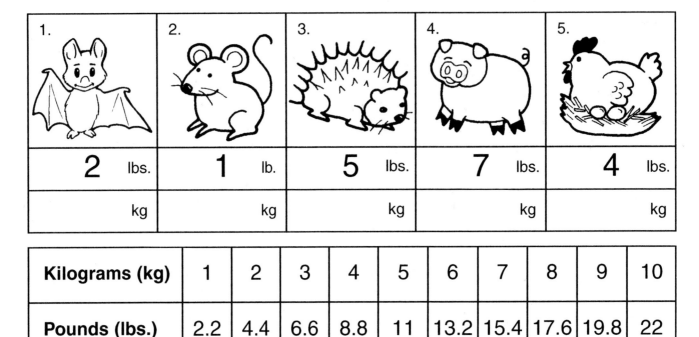

1.	2.	3.	4.	5.
2 lbs.	1 lb.	5 lbs.	7 lbs.	4 lbs.
kg	kg	kg	kg	kg

Kilograms (kg)	1	2	3	4	5	6	7	8	9	10
Pounds (lbs.)	2.2	4.4	6.6	8.8	11	13.2	15.4	17.6	19.8	22

Directions: Each animal's weight is shown in kilograms (kg). Figure out each animal's weight in pounds (lbs.) and write the answer in the box. Use the above chart to help you.

6.	7.	8.	9.	10.
lbs.	lbs.	lbs.	lbs.	lbs.
3 kg	6 kg	9 kg	10 kg	8 kg

Learning Notes

In this unit, children will practice telling time to the hour and to the minute. They will use the information on a calendar to answer questions. In addition, they will put the months of the year in order and will use the information on a calendar to answer questions.

Materials

- calculator
- paper plates
- construction paper
- paper fasteners
- index cards labeled with the days of the week and months of the year

Teaching the Lesson

Before introducing the practice pages, have the children each make his or her own paper plate clock. (See the directions below.)

Directions for making a paper plate clock:

1. Write the numbers 1–12 on the paper plate. Make sure to carefully place the numbers in their correct positions. In addition, mark the lines on the clock for minutes.
2. Cut the hour and minute hands out of construction paper. Make sure the hands are distinctly different in length.
3. Poke a paper fastener through both hands of the clock and the paper plate.

Go over the features of the clock (hour hand, minute hand, face, numbers, etc.). Explain how to figure out the number of minutes by counting the numbers on the face of the clock by 5's. Have the children practice making different times on the clocks.

Before introducing the calendar practice pages, have the children practice placing in order the cards labeled with the days of the week and months of the year. Also have the children answer questions about the cards, such as: "How many days in one week?", "What day comes before/after Thursday?", "How many months are in a year?", "What month is first/last?", etc.

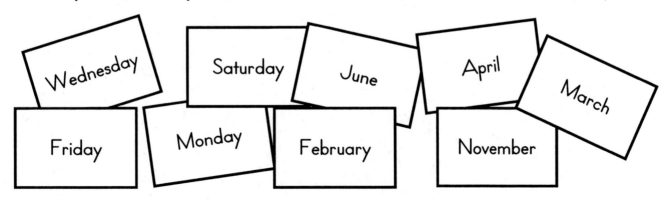

A clock has a "face." The face is numbered from 1 to 12. The numbers are used to tell both the hour and the minutes. There are two hands on a clock. The short hand is called the "hour hand." The long hand is called the "minute hand."

The hour hand points to the hour. As the minute hand moves, so does the hour hand. It takes 60 minutes for the hour hand to reach the next number on the face of the clock. The minute hand tells the number of minutes in the hour. To figure out the number of minutes, count the numbers on the face of the clock by 5's.

hour hand
points to
the hour

minute hand
points to
the minute

Directions: Count the numbers on the face of the clock by 5's. Write the numbers in each box. Make sure to follow the numbers in order. The first two numbers have already been completed for you. Then answer the questions.

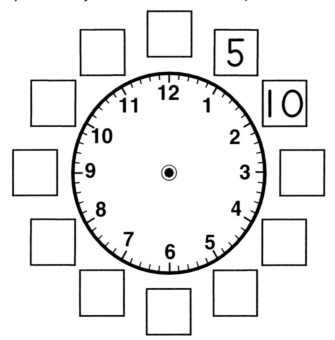

To what number does the minute hand need to point to show:

1. 10 minutes? _____

2. 35 minutes? _____

3. 40 minutes? _____

What two numbers does the minute hand have to be between to show:

4. 8 minutes? _____ _____

5. 23 minutes? _____ _____

6. 58 minutes? _____ _____

7. How many minutes are there in one hour? _____

Directions: Write the correct time.

8.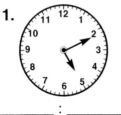

_____ : _____

9.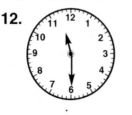

_____ : _____

10.

_____ : _____

11.

_____ : _____

12.

_____ : _____

13.

_____ : _____

Directions: Time can also be measured in days, weeks, and months. Use the calendar to answer the questions.

February 2000

Sunday	Monday	Tuesday	Wednesday	Thursday	Friday	Saturday
		1 Groundhog Day	2	3	4	5
6	7	8	9	10	11	12
13	14 Valentine's Day	15	16	17	18	19
20	21 Presidents' Day	22	23	24	25	26
27	28	29 Leap Day!				

1. What is the name of this month? _____

2. What is the year? _____

3. How many days are in this month? _____

4. How many days does this month usually have? _____

5. Why does this month have an extra day? _____

6. What are the four special days indicated in this month? _____

_____

7. What month comes before this month? _____

8. What month comes after this month? _____

9. How many days are there in one week? _____

10. How many weeks are there in this month? _____

Challenge

11. How many hours are in one week? _____

12. How many hours are in this month? _____

7 Practice •••••••••••••••••• Measuring Time in Months of the Year

January

Sunday	Monday	Tuesday	Wednesday	Thursday	Friday	Saturday
						1
2	3	4	5	6	7	8
9	10	11	12	13	14	15
16	17	18	19	20	21	22
23/30	24/31	25	26	27	28	29

February

Sunday	Monday	Tuesday	Wednesday	Thursday	Friday	Saturday
		1	2	3	4	5
6	7	8	9	10	11	12
13	14	15	16	17	18	19
20	21	22	23	24	25	26
27	28	29				

March

Sunday	Monday	Tuesday	Wednesday	Thursday	Friday	Saturday
		1	2	3	4	
5	6	7	8	9	10	11
12	13	14	15	16	17	18
19	20	21	22	23	24	25
26	27	28	29	30	31	

April

Sunday	Monday	Tuesday	Wednesday	Thursday	Friday	Saturday
						1
2	3	4	5	6	7	8
9	10	11	12	13	14	15
16	17	18	19	20	21	22
23/30	24	25	26	27	28	29

May

Sunday	Monday	Tuesday	Wednesday	Thursday	Friday	Saturday
	1	2	3	4	5	6
7	8	9	10	11	12	13
14	15	16	17	18	19	20
21	22	23	24	25	26	27
28	29	30	31			

June

Sunday	Monday	Tuesday	Wednesday	Thursday	Friday	Saturday
				1	2	3
4	5	6	7	8	9	10
11	12	13	14	15	16	17
18	19	20	21	22	23	24
25	26	27	28	29	30	

July

Sunday	Monday	Tuesday	Wednesday	Thursday	Friday	Saturday
						1
2	3	4	5	6	7	8
9	10	11	12	13	14	15
16	17	18	19	20	21	22
23/30	24/31	25	26	27	28	29

August

Sunday	Monday	Tuesday	Wednesday	Thursday	Friday	Saturday
		1	2	3	4	5
6	7	8	9	10	11	12
13	14	15	16	17	18	19
20	21	22	23	24	25	26
27	28	29	30	31		

September

Sunday	Monday	Tuesday	Wednesday	Thursday	Friday	Saturday
					1	2
3	4	5	6	7	8	9
10	11	12	13	14	15	16
17	18	19	20	21	22	23
24	25	26	27	28	29	30

October

Sunday	Monday	Tuesday	Wednesday	Thursday	Friday	Saturday
1	2	3	4	5	6	7
8	9	10	11	12	13	14
15	16	17	18	19	20	21
22	23	24	25	26	27	28
29	30	31				

November

Sunday	Monday	Tuesday	Wednesday	Thursday	Friday	Saturday
		1	2	3	4	
5	6	7	8	9	10	11
12	13	14	15	16	17	18
19	20	21	22	23	24	25
26	27	28	29	30		

December

Sunday	Monday	Tuesday	Wednesday	Thursday	Friday	Saturday
					1	2
3	4	5	6	7	8	9
10	11	12	13	14	15	16
17	18	19	20	21	22	23
24/31	25	26	27	28	29	30

Directions: Use the calendars above to help you write the exact date for each of the following holidays.

1. Patriot's Day (3rd Monday in April): _____

2. Sadie Hawkins' Day (1st Saturday in November): _____

3. Labor Day (1st Monday in September): _____

4. National Grandparent's Day (1st Sunday after Labor Day): _____

5. Children's Day (2nd Sunday in June): _____

6. Thanksgiving (4th Thursday in November): _____

7. Memorial Day (last Monday in May): _____

8. Father's Day (3rd Sunday in June): _____

9. Mother's Day (2nd Sunday in May): _____

10. Arbor Day (last Friday in April): _____

Learning Notes

In this unit, children will read a thermometer and chart the temperature for a week. They will use the completed chart to answer questions. The children will also use a calculator to convert Farenheit (° F) to Celsius (° C) and to convert Celsius (° C) to Farenheit (° F).

Materials

- calculators
- paper thermometers
- real thermometer
- scratch paper
- local newspaper

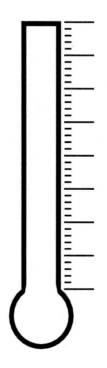

Teaching the Lesson

Go over the markings on the thermometer with the students. Demonstrate how to read the thermometer accurately by holding the thermometer at eye level.

Have the children make their own thermometers out of construction paper. (You may use the thermometer pattern on the right.) Using the paper thermometers, have the children practice making and reading different temperatures.

For a bonus activity, bring in the local newspaper for a week for the children to use in charting each day's temperature.

Have the children familiarize themselves with the buttons on the calculator and how to input information into the calculator. Model how to use the calculators to convert °F to °C and °C to °F.

Converting °F to °C	Converting °C to °F
Step 1: Take the temperature in °F and subtract 32°	**Step 1:** Take the temperature in °C and multiply by 1.8
Step 2: Take the answer and divide by 1.8	**Step 2:** Take the answer and add 32°
Step 3: Record the temperature in °C	**Step 3:** Record the temperature in °F
Example: 87° F − 32° = 55° ÷ 1.8 = 30.5° C	**Example:** 32° x 1.8 = 57.6° + 32° = 89.6° F

A *thermometer* tells the temperature of an object, animal, person, or the weather. To accurately read a thermometer, stand directly in front of the thermometer and make sure that the thermometer is at "eye level." The line in the middle of the thermometer stops at the object's temperature.

Directions: Read the temperature on each thermometer. Write the temperature on the line below.

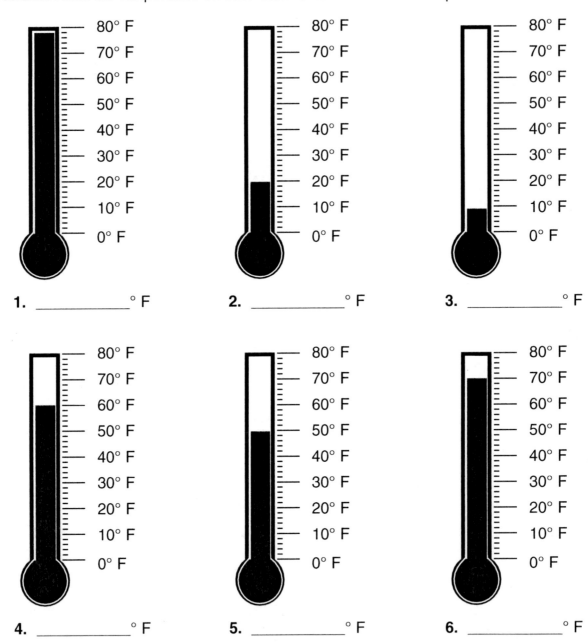

1. _____ ° F

2. _____ ° F

3. _____ ° F

4. _____ ° F

5. _____ ° F

6. _____ ° F

7. Write the temperatures above in order from coldest to the hottest.

_____, _____, _____, _____, _____, _____

Challenge

Be a weather watcher! Make a chart showing each day's high and low temperatures. Is there a pattern to the weather?

Directions: Chart the following temperatures on the weather chart. The temperature for Sunday has already been charted for you.

Sunday: 96° F	Monday: 91° F	Tuesday: 88° F
Wednesday: 100° F	Thursday: 97° F	Friday: 101° F
	Saturday: 87° F	

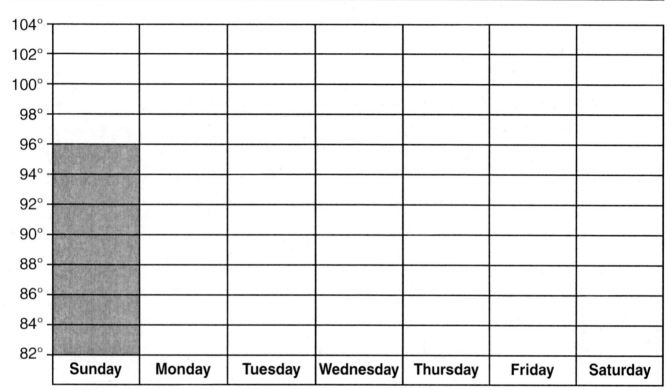

1. What was the coolest day and temperature?_____

2. What was the hottest day and temperature? _____

3. What was the difference in temperature between Friday and Saturday? _____

4. What was the difference in temperature between Tuesday and Wednesday?_____

5. Use a calculator to figure out the average temperature for the week.

   First, add all of the temperatures for the week together.

   _____ + _____ + _____ + _____ + _____ + _____ + _____ = _____

   Next, divide the total temperature for the week by 7 (the number of days in a week).

   _____ ÷ 7 = _____ (average temperature for the week)

   The average temperature for the week is _____.

6. What do you notice about the average temperature for the week and the daily temperatures for the week? _____

_____

8 Practice •••••••••••••• Fahrenheit and Celsius

Fahrenheit and *Celsius* are two ways to measure the temperature in objects, food, animals, people, or weather. People in the United States report the temperature in Fahrenheit (° F). The metric system measures temperature in Celsius (° C).

To convert the temperature in ° F to ° C, the formula is

___° F – 32° F = ___ ÷ 1.8 = ___° C

For example, to convert 87° F to Celsius, take the temperature 87° F – 32° F = 55° F. Then take the answer 55° F and divide by 1.8—the answer is 30.5° C

To convert the temperature in ° C to ° F, the formula is

___° C x 1.8 = ___° C + 32 = ___° F

For example, to convert 32° C to Fahrenheit, take the temperature 32° C x 1.8 = 57.6 ° C. Then take the answer 57.6° C + 32° = 89.6° F—the answer is 89.6° F

Table of Temperature Equivalents

Fahrenheit (° F)	32° F	37° F	41° F	47° F	52° F	57° F	62° F	67° F	72° F	77° F
Celsius (° C)	0° C	2.77° C	5° C	8.33° C	11.11° C	13.88° C	16.66° C	19.44° C	22.22° C	25° C

Directions: Use a calculator or the above chart to help you do the following activities. Remember to convert the temperatures to the same measurement.

Circle the colder temperature.

1. 67° F 25° C

2. 32° F 5° C

3. 77° F 22.22° C

4. 47° F 2.77° C

5. 41° F 0° C

Circle the warmer temperature.

6. 37° F 11.11° C

7. 72° F 8.33° C

8. 52° F 13.88° C

9. 62° F 19.44° C

10. 57° F 16.66° C

Learning Notes

In this unit, children will learn to find the area for different shapes in square units, square centimeters, and square inches. They will also find the perimeter for different shapes.

Materials

- multilink cubes

Teaching the Lesson

Make any flat shape using the multilink cubes. Ask the children to find the area by counting the number of cubes used.

There are 24 cubes in the shape.
The area is 24 square units.

1	2	3	4	5	6
7	8	9	10	11	12
13	14	15	16	17	18
19	20	21	22	23	24

To find the perimeter, have the children count the number of cubes on each side of the shape.

$4 + 6 + 4 + 6 = 20$

The perimeter is 20 units.

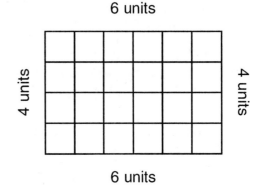

6 units

4 units   4 units

6 units

Extension

Using the multilink cubes, make a cube shape. Have the children find the area for one side of the cube. (The children might need to break apart the cube in order to count all of the multilink cubes used to make one side of the shape.)

Have the children practice making a variety of flat shapes to find the area and the perimeter and also a variety of cubes to find the total area.

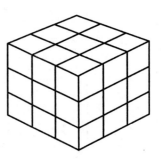

The area for one side is 9 square units.

The *area* is the number of square units (inches, centimeters, feet, miles, meters, etc.) a shape covers. To find the area, count the number of units in the shape. Look at the example below.

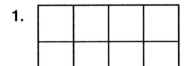

1 cm ☐ 1 cm = 1 square centimeter

Each side of the square is 1 cm long.

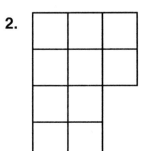

There are 5 square centimeter units in the shape.

The area is 5 square centimeter units.

Directions: Find the area of each shape. Each box represents a square centimeter. Write the answers on the lines.

1.

The area is

_____ square cm.

2.

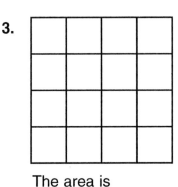

The area is

_____ square cm.

3.

The area is

_____ square cm.

Directions: Color in the squares to make a shape that is the correct number of square centimeters. Colored squares must touch sides of other colored squares to make one shape.

4. 10 square centimeters

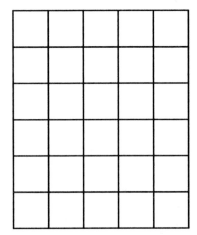

5. 7 square centimeters

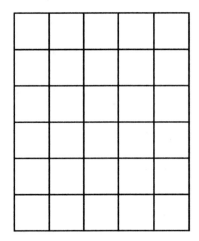

6. 15 square centimeters

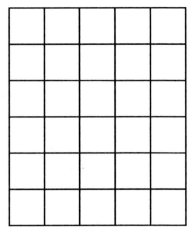

The *perimeter* is the outside measurement of a shape. To find the perimeter, count the number of cubes on each side of the shape. Then add all the sides together.

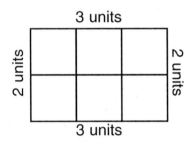

$2 + 3 + 2 + 3 = 10$

The perimeter is 10 units.

Directions: Find the perimeter for each shape. Use the multilink cubes to help you.

1.

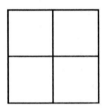

The perimeter is
_____ units.

2.

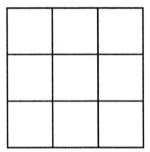

The perimeter is
_____ units.

3.

The perimeter is
_____ units.

4.

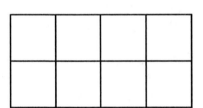

The perimeter is
_____ units.

5.

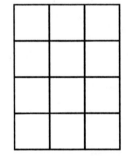

The perimeter is
_____ units.

6.

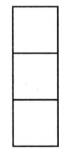

The perimeter is
_____ units.

7.

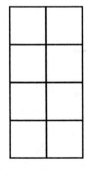

The perimeter is
_____ units.

8.

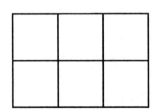

The perimeter is
_____ units.

9.

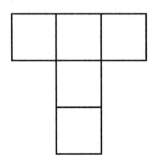

The perimeter is
_____ units.

•••• **Finding the Perimeter and the Area**

The *perimeter* is the outside measurement of a shape. 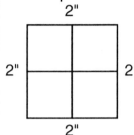 Each block represents a 1" square. Each side is 2" long. 2 + 2 + 2 + 2 = 8 The perimeter is 8 inches.	The *area* is the inside measurement of a shape. Count the blocks. 1 + 1 + 1 + 1 = 4 The area is 4 square inches.

Directions: Find the perimeter and area for each shape. Remember, each block represents a 1" square. Remember to label the perimeter in inches and the area in square inches.

1.

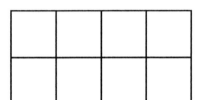

Perimeter:_____

Area:_____

2.

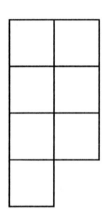

Perimeter:_____

Area:_____

3.

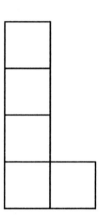

Perimeter:_____

Area:_____

4.

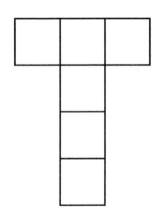

Perimeter:_____

Area:_____

5.

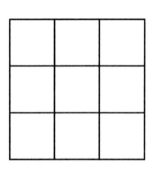

Perimeter:_____

Area:_____

6.

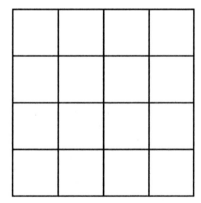

Perimeter:_____

Area:_____

10 Problem Solving

Arabic and Roman Numerals

Arabic and Roman numerals are ways of writing numbers. Arabic numerals are used throughout the world and look like this: 1, 2, 3, 60, 71, etc. Roman numerals are also used throughout the world and are often used to show the year that a movie was made or a book was published. Roman numerals look like this: I, V, X, L, C, etc.

Arabic	1	2	3	4	5	6	7	8	9
Roman	I	II	III	IV	V	VI	VII	VIII	IX

Arabic	10	11	12	13	14	15	16	17	18	19
Roman	X	XI	XII	XIII	XIV	XV	XVI	XVII	XVIII	XIX

Arabic	20	30	40	50	60	70	80	90	100
Roman	XX	XXX	XL	L	LX	LXX	LXXX	XC	C

Arabic	150	200	300	400	500	600	700	800	900	1000
Roman	CL	CC	CCC	CD	D	DC	DCC	DCCC	CM	M

Directions: Write the following numbers using Roman numerals. Use the chart above to help you.

1. 27 ________________

2. 75 ________________

3. 890 ________________

4. 950 ________________

5. 2000 ________________

6. 650 ________________

7. 62 ________________

8. 56 ________________

9. 110 ________________

10. 370 ________________

11. 98 ________________

12. 240 ________________

Challenge: Write the year you were born in Arabic numerals and then in Roman numerals.

Arabic ____________________ Roman ____________________

Miles (mi.)	1	2	3	4	5	6	7	8	9	10
Kilometers (km)	1.6	3.2	4.8	6.4	8.0	9.6	11.2	12.8	14.4	16.0

Directions: Solve each word problem. Show your work on a separate sheet of paper.

1. Lauren rode her bike 3 miles to the grocery store and 2 miles to her friend's house. How many kilometers did Lauren ride in all? Lauren rode her bike _____ km.

2. Bobby lives 4 miles from school. He takes the school bus to and from school. How many kilometers does the bus travel? The bus travels _____ km.

3. Celeste's grandmother lives 5 miles away. How many kilometers does Celeste travel to visit her grandmother and return home? Celeste travels _____ km.

4. Vince skated 1 mile to Joshua's house then 6 more miles to Devin's house. How many kilometers did Vince skate? Vince skated _____ km.

Kilometers (km)	1	2	3	4	5	6	7	8	9	10
Miles (mi.)	.62	1.24	1.86	2.48	3.10	3.72	4.34	4.96	5.58	6.20

5. George's scooter can travel 8 kilometers in an hour. How many miles can his scooter go in one hour? George's scooter can go _____ miles in one hour.

6. Vivian drove 3 kilometers to the mall and 6 kilometers to the grocery store. How many miles did Vivian drive? Vivian drove _____ miles.

7. Henry goes to the park every day. The park is 2 kilometers away. How many miles does Henry travel going to and from the park? Henry travels _____ miles.

8. Stephanie rode her skateboard to the bakery and then to the movie theater. She skateboarded 7 kilometers in all. How many miles did she ride her skateboard? Stephanie skateboarded _____ miles.

In the Chinese calendar, each year is designated by one of 12 animals.

Rat	Ox	Tiger	Rabbit	Dragon	Snake
1924	1925	1926	1927	1928	1929
1936	1937	1938	1939	1940	1941
1948	1949	1950	1951	1952	1953
1960	1961	1962	1963	1964	1965
1972	1973	1974	1975	1976	1977
1984	1985	1986	1987	1988	1989
1996	1997	1998	1999	2000	2001
2008	2009	2010	2011	2012	2013

Horse	Sheep	Monkey	Rooster	Dog	Boar
1930	1931	1932	1933	1934	1935
1942	1943	1944	1945	1946	1947
1954	1955	1956	1957	1958	1959
1966	1967	1968	1969	1970	1971
1978	1979	1980	1981	1982	1983
1990	1991	1992	1993	1994	1995
2002	2003	2004	2005	2006	2007
2014	2015	2016	2017	2018	2019

Directions: Use the Chinese calendar to answer the questions.

1. Under what animal sign were you born? _____

2. What is the animal sign for the current year? _____

3. What is the pattern for the animals on the Chinese calendar?

   _____

   _____

4. How would you figure out the animal sign for someone born in the year 1900?

   _____

   _____

5. How would you figure out the animal sign for someone born in the year 2036?

   _____

   _____

6. On a separate piece of paper, make a calendar showing the animal signs for the birthday year of each of your family members. Write two sentences telling about the calendar you made.

Directions: Read the clues below. Then fill in every box on the chart. If the answer is "no," make an "X" in the box. If the answer is "yes," then make an "O" in the box. Fill in the final length and weight of each animal at the bottom of the page. (*Hint:* Each animal has a different length and weight.)

Length

	1 ft.	3 ft.	4 ft.	6 ft.
Cat				
Dog				
Possum				
Rabbit				

Weight

	2 lbs.	5 lbs.	8 lbs.	10 lbs.
Cat				
Dog				
Possum				
Rabbit				

Clues

- Possum weighs 2 pounds.

- Rabbit is 3 feet long.

- Dog weighs 10 pounds.

- Cat does not weigh 5 pounds.

- Possum is not the longest nor the shortest animal.

- Cat is not the heaviest nor the lightest animal.

- Dog is longer than Rabbit.

	Length	Weight			Length	Weight
1. Cat	_____	_____		**3.** Possum	_____	_____
2. Dog	_____	_____		**4.** Rabbit	_____	_____

Take a Trip to a City

> **Note to the Adult Supervisor:** For this activity, children will use the Internet. Before allowing access to the Internet, verify that all necessary permission has been granted for the child to use the Internet.

Suppose you are going to take a trip from Los Angeles, California, to Prescott, Arizona. You are going to start at Los Angeles and stop at Prescott. Use the Internet to help you find the fastest way to get there.

Steps

1. Go to http://www.mapquest.com/

2. Click on the button about driving directions.

3. On the starting address area, type in Los Angeles for the city and CA for the state.

4. For your destination or ending point, type Prescott for the city and AZ for the state.

5. Under the route type or options, select "City-to-City."

6. At the bottom of the page, click the button that calculates or gets your directions.

Measure the Distances

- What distance will you travel on the Hollywood Freeway? _____

- How many miles will you travel east on I-10 to Arizona? _____

- How far will you drive northeast on State Route 71 to State Route 89? _____

- What will be the total distance of this trip? _____

Take a Trip to an Amusement Park

> **Note to the Adult Supervisor:** For this activity, children will use the Internet. Before allowing access to the Internet, verify that all necessary permission has been granted for the child to use the Internet.

Imagine that this summer you are going to visit Disneyland in Anaheim, California. You will then travel to Walt Disney World in Orlando, Florida. Use the Internet to help you find the fastest way to reach your destination.

Steps

1. Go to http://www.mapquest.com/

2. Click on the button about driving directions.

3. On the starting address area, type in Anaheim for the city and CA for the state.

4. For your destination or ending point, select "Amusement Parks" on the dropdown menu. Type Orlando for the city and FL for the state.

5. Under the route type or options, select "City-to-City."

6. At the bottom of the page, click the button that calculates or gets your directions.

Measure the Distances

- How many miles will you travel on State Route 60 to I-10? _____

- How many miles will you travel from I-10 East to Arizona? _____

- How many miles will you travel going I-75 South to Florida's Turnpike? _____

- What is the total distance of your trip? _____

Plan Your Own Trip

- If you had a choice of one place you would like to visit in the United States, where would you go? Why did you choose this place?

- On a separate piece of paper, write the starting point and destination you chose, explain the route you selected, and calculate the total distance of your trip. Use the Web site http://www.mapquest.com/ to plan your trip.

Page 6

1.–6. Answers will vary.

7. Yes, because I would need to use fewer fingers.

Page 7

1.–10. Answers will vary.

Page 8

1.–9. Answers will vary.

10. If my foot was longer than 12 inches, it would take fewer feet to measure items.

Page 10

1.–8. Answers will vary.

Page 11

1.–10. Answers will vary depending upon the measurement tool used.

Page 12

1.–3. Answers will vary.

Page 14

1. 4 inches
2. 3 inches
3. 2 inches
4. 5 inches
5. less
6. less
7. less
8. more
9. less
10. less

Page 15

1. 30 in.; $2\frac{1}{2}$ ft.
2. 6 in.; $\frac{1}{6}$ yd.
3. $1\frac{1}{2}$ ft.; $\frac{1}{2}$ yd.
4. 36 in.; 1 yd.
5. 12 in.; 1 ft.
6. 2 ft.; $\frac{2}{3}$ yd.
7. 24 in.; 2 ft.
8. 18 in.; $\frac{1}{2}$ yd.

Page 16

Number of miles for each state highway:

State Highway 1: 50 miles

State Highway 2: 20 miles

State Highway 3: 60 miles

State Highway 4: 40 miles

State Highway 5: 30 miles

State Highway 6: 100 miles

State Highway 7: 60 miles

State Highway 8: 60 miles

1. 150 miles
2. 170 miles
3. 8 to 5 to 3 to 2
4. 3 to 2 to 1 to 4; or 5 to 6; or 5 to 8 to 7
5. 3 to 2 to 1 to 4 = 170 miles; 5 to 6 = 130 miles; 5 to 8 to 7 = 150 miles

6. 6 to 4; or 5 to 3 to 2 to 1; or 8 to 7 to 4
7. 6 to 4 = 140 miles; 5 to 3 to 2 to 1 = 160 miles; 8 to 7 to 4 = 160 miles
8. 6; or 8 to 7; or 5 to 3 to 2 to 1 to 4
9. 6 = 100 miles; 8 to 7 = 120 miles; 5 to 3 to 2 to 1 to 4 = 200 miles
10. Answers will vary.

Page 18

1. 4 cm
2. 6 cm
3. 3 cm
4. 6 cm
5. 4 cm
6. 8 cm
7. 6 cm
8. 1 cm

Page 19

1. 3 + 3 + 3 = 9 cm
2. 3 + 2 + 5 + 2 = 12 cm
3. 1 + 1 + 1 + 1 + 1 + 1 + 1 + 1 = 8 cm
4. 2 + 2 + 2 + 2 = 8 cm
5. 3 + 3 + 6 + 2 = 14 cm
6. 2 + 6 + 7 = 15 cm

Page 20

1. 9 cm
2. 2 cm

3.–7. Answers will vary.

Page 22

1. gallon—4 quarts
2. 2 quarts—8 cups
3. 2 cups—1 pint
4. 2 pints—1 quart
5. 16 cups
6. 4 pints
7. $\frac{1}{2}$ pint
8. 1 gallon
9. 4 cups
10. 4 quarts
11. $\frac{1}{2}$ gallon
12. 1 quart

Page 24

1. L
2. L
3. L
4. dL
5. dL
6. dL
7. dL
8. dL
9. dL
10. dL
11. L
12. dL
13. dL
14. dL
15. L

Page 26

1. 32
2. 64
3. 96
4. 128
5. 144
6. 3
7. 5
8. 6
9. 7
10. 8
11. 10
12. lb.
13. lb.
14. oz.
15. oz.
16. lb.
17. oz.
18. oz.
19. lb.
20. oz.
21. lb.
22. oz.
23. oz.
24. oz.
25. lb.
26. oz.

Challenge: Answers will vary.

Page 27

1. 3,000
2. 4,000
3. 6,000
4. 7,000
5. 9,000
6. 10,000
7. 1
8. 3
9. 4
10. 5
11. 7
12. 8
13. 9
14. kg
15. g
16. g
17. g
18. kg
19. g
20. g
21. g
22. kg
23. g
24. kg
25. kg
26. kg
27. kg
28. g

Challenge: Answers will vary.

Page 28

1. .90
2. .45
3. 2.25
4. 3.15
5. 1.80
6. 6.6
7. 13.2
8. 19.8
9. 22.0
10. 17.6

Page 30

Numbers on the face of the clock: (5), (10), 15, 20, 25, 30, 35, 40, 45, 50, 55, 60 (or 0)

1. 2
2. 7
3. 8
4. 1 and 2
5. 4 and 5
6. 11 and 12
7. 60 minutes
8. 6:05
9. 6:15
10. 12:00
11. 5:10
12. 11:30
13. 8:45

Page 31

1. February
2. 2000
3. 29
4. 28
5. Because of Leap Day. Every 4 years an extra day is added to the month of February to make up for shortening the 3 previous years by 1/4 of a day each year. (A year is actually 365.25 days long.)
6. Groundhog Day, Valentine's Day, Presidents' Day, and Leap Day
7. January
8. March
9. 7 days
10. 4
11. 168 hours in one week
12. 696 hours in this month

Page 32

1. April 17
2. Nov. 4
3. Sept. 4
4. Sept. 10
5. June 11
6. Nov. 23
7. May 29
8. June 18
9. May 14
10. April 28

Page 34

1. 80° F
2. 20° F
3. 10° F
4. 60° F
5. 50° F
6. 70° F
7. 10° F; 20° F; 50° F; 60° F; 70° F; 80° F

Challenge: Answers will vary.

Page 35

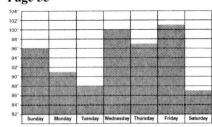

1. Saturday, 87° F
2. Friday, 101° F
3. 101° F – 87° F = 14° F
4. 100° F – 88° F = 12° F
5. 96° F + 91° F + 88° F + 100° F + 97° F + 101° F + 87° F = 660° F; 660 ÷ 7 = 94.29° F; The average temperature for the week is 94° F
6. Answers will vary.

Page 36

1. 67° F
2. 32° F
3. 22.22° C
4. 2.77° C
5. 0° C
6. 11.11° C
7. 72° F
8. 13.88° C
9. 19.44° C
10. 16.66° C

Page 38

1. 8 sq. cm
2. 10 sq. cm
3. 16 sq. cm
4.–6. Shapes will vary.

Page 39

1. 8 units
2. 12 units
3. 6 units
4. 12 units
5. 14 units
6. 8 units
7. 12 units
8. 10 units
9. 12 units

Page 40

1. Perimeter 12 in.; Area 8 sq. in.
2. Perimeter 12 in.; Area 7 sq. in.
3. Perimeter 12 in.; Area 5 sq. in.
4. Perimeter 14 in.; Area 6 sq. in.
5. Perimeter 12 in.; Area 9 sq. in.
6. Perimeter 16 in.; Area 16 sq. in.

Page 41

1. XXVII
2. LXXV
3. DCCCXC
4. CML
5. MM
6. DCL
7. LXII
8. LVI
9. CX
10. CCCLXX
11. XCVIII
12. CCXL

Challenge: Answers will vary.

Page 42

1. 8 km
2. 12.8 km
3. 16 km
4. 11.2 km
5. 4.96 mi.
6. 5.58 mi.
7. 2.48 mi.
8. 4.34 mi.

Page 43

1. Answers will vary.
2. Answers will vary.
3. The animals repeat every 12 years.
4. Start with 1924 and count backwards (on the animals) until the year 1900 is reached. Or start on 1924 and subtract 24 which equals 1900. Twelve can go into 24 two times evenly. So the year is the year of the Rat.
5. Start with 2019 and count ahead to the year 2036. Or add 12 to 2019 and count 5 more animals to the year of the Dragon.
6. Answers will vary.

Page 44

Length

	1 ft.	3 ft.	4 ft.	6 ft.
Cat	O	X	X	X
Dog	X	X	X	O
Possum	X	X	O	X
Rabbit	X	O	X	X

Weight

	2 lbs.	5 lbs.	8 lbs.	10 lbs.
Cat	X	X	O	X
Dog	X	X	X	O
Possum	O	X	X	X
Rabbit	X	O	X	X

1. Cat: 1 ft., 8 lbs.
2. Dog: 6 ft., 10 lbs.
3. Possum: 4 ft., 2 lbs.
4. Rabbit: 3 ft., 5 lbs.

Page 45

1.5 miles (2.4 km)
174.2 miles (280.3 km)
24.3 miles (39.1 km)
380.6 miles (612.4 km)

Page 46

18.0 miles (29.0 km)
153.3 miles (246.7 km)
109 miles (175.4 km)
2519.6 miles (4054 km)